FOREST PHARMACY
Medicinal Plants In American Forests

by
STEVEN FOSTER

Forest History Society
Durham, North Carolina
1995

The Forest History Society is a nonprofit, educational institution dedicated to the advancement of historical understanding of human interaction with the forest environ-ment. The Society was established in 1946. Interpretations and conclusions in FHS publications are those of the authors; the Society takes responsibil-ity for the selection of topics, the competence of the authors, and their freedom of inquiry.

Forest History Society
701 Vickers Avenue
Durham, North Carolina 27701
(919) 682–9319

This book is a history of the usage of medicinal plants in America and is not intended as a guide to self-diagnosis or treatment.

Library of Congress Cataloging-in-Publication Data

Foster, Steven, 1957–
    Forest pharmacy : medicinal plants in American forests / by Steven Foster.
        p.   cm. — (Forest History Society issues series)
    Includes bibliographical references.
    ISBN 0-89030-051-8
    1. Materia medica, Vegetable—United States—History.
    2. Medicinal plants—United States—History.   I. Title.
    II. Series.
    RS 164.F72   1995
    615'.32'0973—dc20   95-14291
                                                                                    CIP

On the cover—Americans have long used and continue to use forest plants as powerful sources of medicine. The photo shows three medicinal plants found in the forests of the Wasatch Mountains in Utah: penstemon, yellow sweet clover, and wild flax. Photo by Steven Foster.

Title page and page 1—Pen-and-ink drawing of Pacific yew by Sydney Rust.

Page 15—Engraving of sassafras from volume 2 of *American Medical Botany, Being a Collection of the Native Medicinal Plants of the United States . . .*, by Jacob Bigelow (Boston, Massachusetts: Cummings and Hilliard, 1818), plate 35.

Page 24—Engraving of dogwood from volume 2 of *American Medical Botany, Being a Collection of the Native Medicinal Plants of the United States . . .*, by Jacob Bigelow (Boston, Massachusetts: Cum-mings and Hilliard, 1818), plate 28.

Page 29—Engraving of tulip-tree from volume 2 of *American Medical Botany, Being a Collection of the Native Medicinal Plants of the United States . . .*, by Jacob Bigelow (Boston, Massachusetts: Cum-mings and Hilliard, 1818), plate 31.

Forest History Society Issues Series

The Forest History Society was founded in 1946. Since that time, the Society, through its research, reference, and publication programs, has advanced forest and conservation history scholarship. At the same time, it has translated that scholarship into formats useful for people with policy and management responsibilities. For nearly five decades the Society has worked to demonstrate history's significant utility.

The Forest History Society Issue Series is the latest and most explicit contribution to history's utility. With guidance from the Advisory Committee, the Society selects issues of importance today that also have significant historical dimensions. Then we invite authors of demonstrated knowledge to examine an issue and synthesize its substantial literature, while keeping the general reader in mind.

The final and most important step is making these authoritative overviews available. Toward that end, each pamphlet is distributed to people with management, policy, or legislative responsibilities who will benefit from a deepened understanding of how a particular issue began and evolved.

The Issues Series—like its Forest History Society sponsor—is nonadvocacy. The series aims to present a balanced rendition of often contentious issues. While all views are aired, the focus is on consensus. The pages that follow trace the history of American plant medicines, anchoring the discussion in a European and Asian context.

The Society gratefully acknowledges financial support from the Huss Foundation for this third title in the Issues Series.

Advisory Committee

Contents

Tables

Photos

Overview

- Medicinal plants have always played an important role in medicine. More people throughout the world rely on herbal remedies than rely on orthodox western medicine.

- Approximately 25 percent of prescription drugs contain at least one plant-derived compound, or are based on plant-derived chemical models. That percentage has remained relatively stable (± 1 percent) since 1959. Despite this successful continuity, American researchers conduct very little research on American medicinal plants.

- Dramatic technological advances in molecular biology and laboratory automation have helped spark renewed interest in plants as sources of novel compounds for new drugs.

- Of 21,757 species of flowering plants in North America (not including Mexico), there is documentation that Native American groups used at least 2,147 (10 percent) for medicinal purposes.

- Attempts to introduce American medicinal plants into medicine and pharmacy in the late-eighteenth century resulted in over 130 indigenous plant species being admitted to the *United States Pharmacopoeia* and/or the *National Formulary*. Most of those species are still traded as botanicals (medicinal, fragrant, or flavor materials) in international markets. The majority are wild-harvested from forest habitats, chiefly in the southern Appalachian Mountains.

- In a number of developed western nations, notably Germany, many of those same species carry therapeutic claims and are sold in pharmacies. These plant species are available in the United States, primarily in health and natural food markets, in the form of dietary supplements that may not legally carry or imply health claims on their labels.

- Economic and regulatory disincentives for American medicinal plants in the United States discourage scientific research. Therefore, most modern scientific research on American medicinal forest species is conducted in Europe or Asia. A new regulatory mechanism, allowing for rational, acceptable medicinal claims about American forest species might encourage more research.

- Demand for medicinal plants is increasing in international markets, indicating the need for assuring sustainable sources of supply, developing commercial cultivation methods, and implementing quality assurance and identity standards.

- Growing interest in medicinal plant uses is stimulating research aimed at verifying historic therapeutic claims of safety and efficacy as well as discovery of new drugs.

Medicinal Plants: A Modern Renaissance

There are about a quarter million species of known flowering plants in the world, and people have used about a quarter of these at some time for medicinal purposes. Plant medicines are widely used today all over the world, both in developed and developing countries. How they are used and how they are developed varies widely. In the United States, scientists usually develop new drugs by isolating the chemical components of a plant and then basing the structure of the new drug on that chemical model. One drug that provides a good example of how plant medicinals have been developed in the United States is the successful anticancer drug taxol. (Taxol is the common name used to refer to the generic drug paclitaxel, and is the registered trade name [Taxol®] of Bristol-Myers Squibb Company in Princeton, New Jersey.) Partly due to taxol's success, interest in medicinal plant research has heightened in recent years.

Taxol as an Example of Plant-Derived Medicines

There are many problems inherent in developing plant-derived medicines, but there is also much promise. Taxol, more than most plant-derived medicines, exemplifies both the promise and the problems. The original source for taxol is the bark of a relatively neglected though not rare tree, the Pacific yew.* Once scientists discovered taxol's unique mechanism and then ran successful clinical studies, the need for a sustainable supply of the

*The scientific names for medicinal plants mentioned in this book are listed in table 4, p. 55.

raw material became clear. From that point, the story of taxol and the Pacific yew is intertwined with human use and exploitation of natural resources, providing an example of how those resources, including the habitats in which they occur, can be sustained.

For over thirty years, the National Cancer Institute (NCI) has been at the forefront of clinical evaluation of drugs approved for cancer treatment. It has evaluated drugs including those derived from mayapple for the treatment of small-cell lung cancer and cancer of the testicles; and drugs derived from vincristine and vinblastine, alkaloids from the Madagascar periwinkle used in chemotherapy for Hodgkin's disease, various lymphomas, Wilm's tumor, and other cancers. The discovery of taxol resulted from the NCI's general plant screening program, undertaken from 1958 to 1980, in which it tested over thirty-five thousand plant species for anticancer activity. Taxol emerged as one of the most important new anticancer drugs of the twentieth century. But in 1980 the federal government eliminated the plant screening program, citing the program's expense and the disappointingly few new drugs that resulted from it; eight years later, partly due to the success of taxol as a new chemotherapeutic agent, the program started up again.

The foundation for taxol's discovery was laid in 1961 when NCI gave Monroe Wall and his team at the Research Triangle Institute the first contract to isolate active plant-derived agents. The following year, as part of the exploratory plant screening program, the United States Department of Agriculture (USDA) collected Pacific yew bark for the NCI. An assay in 1964 confirmed toxicity of taxol to certain cancer cell lines, igniting researchers' interest. Taxol was isolated in 1969, and by 1971 its structure had been determined. In 1977 the NCI started preclinical development of taxol; tests showed that the compound had significant activity in a number of human tumor test systems, including breast cancer. In 1979 researchers at the Albert Einstein Institute discovered taxol's unique mechanism of action. By 1980 toxicological studies had begun and formulation studies were completed. This illustrates how the time between discovery of a new plant drug and its entry into the marketplace can be, and often is, up to twenty years.

But taxol still had to take the long road to clinical success, which began in 1983 when tests evaluating safety to humans and alternative doses and regimes (called phase 1 clinical trials) were approved. Trials designed to evaluate taxol's effectiveness in various cancer types (early-phase II clinical trials) developed slowly, due in part to severe supply shortages of the rare taxol compound, which had to be extracted from Pacific yew tree bark. At the time, the NCI conducted all taxol research, and it supplied taxol to

Pacific yew is a small, understory tree in pine and fir forests that mainly grows through the Pacific Northwest to Alaska. It is the subject of intense interest because the tree's bark first provided a source for the drug paclitaxel, a promising anticancer chemotherapy drug for treating ovarian and breast cancers. The Pacific yew is today cultivated so that its leaves can be used to make taxol; one forest products company had planted more than four million yew trees by 1993 and planned to cultivate ten million more by 1996. Photo courtesy of Weyerhaeuser Co.

Pacific yew

physicians in dozens of treatment referral centers, which were partners with the NCI in conducting clinical trials.

When preliminary reports on taxol studies showed an unexpectedly high response rate of 30 percent in ovarian cancer, it sparked excitement and increased demand for taxol, compounding the supply problem. Studies done in 1991 and 1992 showed that women who had breast cancer, which was responsible for the death of forty thousand women per year, responded positively to the drug. Positive results were also recorded in patients with advanced malignancy, including lung cancer, cancer of the head and neck region, malignant melanoma, and lymphomas. With these results, the need to develop a sustainable source of the compound grew more acute. In January 1993 taxol was approved as a therapeutic agent for refractory ovarian cancer; in April 1994 it was also approved for certain forms of breast cancer.

Supply is often a limiting factor, as it was with taxol, in expanding preliminary laboratory discoveries as well as in establishing clinical development of new natural products. The Pacific yew is a small evergreen tree between ten and forty (rarely up to seventy) feet tall. It is found throughout northwest forests from southeast Alaska to northern California, eastward to Montana and Idaho. Although rare in pure stands and not common, the tree is sufficiently abundant not to be considered rare or endangered. Taxol is one of over a dozen chemical compounds, known as taxanes, identified from various species of yew.

The genus *Taxus* is represented by twenty-three species in northern temperate climates. Commonly known as yews, they are among the most widely planted ornamental shrubs, extensively used in landscaping. English yew is the most well-known in American horticulture with more than two hundred fifty cultivated varieties. Japanese yew, introduced to the United States in 1862, is also widely planted.

Yew's medicinal use did not begin with the discovery of taxol. Native Americans traditionally treated with yew a wide range of ailments, such as rheumatism, tuberculosis, gonorrhea, and wounds. Old World cultures also used yew toxins on poison arrow tips to kill both fish and animals. Mixed with clarified butter, yew pitch was used to treat cancer, portending its use by modern western societies.

As early shortages of taxol demonstrated, however, the slow-growing Pacific yew simply could not supply the bark that would be demanded if taxol became a successful cancer treatment. Although the yew is a renewable resource, destructive bark harvesting techniques and the tree's relative scarcity meant that other strategies for producing taxol would have to be developed. By the late-1980s, vigorous programs began worldwide to find alternative taxol sources and finally in a breakthrough French scientists developed a semisynthetic process that uses compounds found in every yew species to create taxol in the laboratory. This method of taxol semisynthesis uses yew needles and provides a sustainable source of supply, so Pacific yew bark no longer serves as the sole source. Needles can be harvested continuously, unlike bark, which can be harvested only once. The French process permits large-scale yew cultivation and harvesting of the rapidly renewable needle, helping to ensure a sustained supply.

Since the time when the semisynthetic process was developed, taxol has been completely synthesized. But it is still less expensive either to extract the compound from natural sources or to produce it by semisynthesis (using natural compounds) rather than to synthesize it completely. For taxol, as for many other compounds, synthetic commercial production is simply not economically feasible.

Before scientists developed the process using needles, it took about thirty pounds of bark to produce one gram of taxol. From 1976 to 1985, annual demand for Pacific yew bark ranged from two thousand to fifteen thousand pounds and the total yield from all this bark was only three pounds of taxol. But because of the successful early clinical trials, demand for Pacific yew bark soared to sixty thousand pounds in 1987–88. Demand for sixty thousand more pounds of the bark in 1989 threatened the tree's survival.

English yew, the most well-known yew in American horticulture, is represented by more than two hundred fifty cultivars. In the late-1980s, as the need for larger quantities of taxol increased, various programs were developed to search for sources other than yew bark. Today a component derived from the needles of English yew provides a semisynthetic precursor used in manufacturing taxol. Photo by Steven Foster.

English yew

In 1991 the NCI began working with Bristol-Myers Squibb (BMS), a pharmaceutical company, to develop taxol production and alternative taxol sources. BMS undertook an inventory of Pacific yew on government lands, and another company began to harvest bark from land designated for clearcutting. In August 1992 the Pacific Yew Act established an Interagency Yew Committee to develop a rigorous Pacific yew harvesting and management program. In 1991 over 1.6 million pounds of the bark was harvested annually, producing 130 kg of taxol; by 1993 the company produced 230 kg of taxol.

One forest products company, Weyerhaeuser, is now cultivating yew in order to use the leaves as a source of taxol. Breeding selectively for desired traits, environmental variables, and control of undesirable traits affecting taxol purity, the company's research is geared toward optimizing taxane production. Weyerhaeuser started with two hundred thousand Pacific yew seedlings in 1991. By 1993 it had four million plants in the ground and a target planting of ten million seedlings over the following two years. The company also developed mechanical harvesting techniques. Dozens of other private companies, government agencies, and university research groups have increased their research efforts to produce sources of taxanes that can be converted into taxol. Success with taxol has sparked interest generally in natural products research in the United States.

The NCI, through its extensive efforts to respond to the taxol supply crisis, learned many important lessons, especially the need for close communication among organizations responsible for drug procurement and clinical investigations. As a result of this experience, the agency developed a strategy to initiate exploratory research programs for a large scale-up of raw material production as soon as proof of a compound's antitumor activity is confirmed.

The taxol story focuses renewed interest in medicinal plant research in the United States and throughout the world. It also focuses attention on the problems inherent in obtaining a sustainable source for compounds derived from scarce or rare plants, and on the pressures that demands for supply place on plant populations.

Medicinal Plants in the Modern World

About one-quarter of the more than quarter million known species of flowering plants have been used at some time for medicinal purposes. At least eighty thousand plant species are documented as traditional medicines worldwide. But plant medicines are not historical relics. They have significant importance today both in developed and developing countries. The 1985 market value of prescription and over-the-counter plant-based drugs in developed countries was an estimated $43 billion. Twenty-five percent of prescription drugs dispensed in the United States contain one or more active constituents obtained from higher (flowering) plants. In addition to anticancer drugs, the list includes compounds from foxglove, used in the management and treatment of heart disease, and atropine, a toxic alkaloid from belladonna used to reduce salivary secretions and to control bronchial secretions during surgery. Curare or South American arrow poison, extracted from the bark of several South American tree species, is used to relax skeletal muscles during surgery. The famous pain killers morphine and codeine are narcotic and habit-forming alkaloids derived from the opium poppy.

One hundred nineteen distinct chemical substances derived from ninety-one plant species are used in orthodox western medicine. Of these, eighty-eight (74 percent) were adopted after scientists investigated the traditional or folk uses. These figures refer only to chemical compounds extracted from plants used in modern western medicine. Globally, an even greater number of plant species are used safely and effectively to prevent or treat disease. As much as 80 percent of the world's population relies on traditional medicine, chiefly herbal remedies.

About one-quarter of the more than quarter million known species of flowering plants have been used at some time for medicinal purposes. Foxglove, a biennial plant whose dried leaves are a source of highly toxic glycosides, is among them. It is used in modern medicine to treat congestive heart failure and also to elevate the blood pressure in patients with weak hearts. Found in New England, it can be lethal. Photo by Steven Foster.

Foxglove

Medicinal Plants in the United States

The use of medicinal plants takes several forms. In the United States, scientists primarily develop new drugs by basing their formulation on isolated single chemical components of a plant that can be patent protected. Patent protection is important because it equates to profitability in new-drug development. Pharmaceutical companies limit medicinal plant research to discovering single active compounds that can be used to develop new prescription drugs for a specific chronic disease, such as cancer, heart disease, or diabetes. This is the so-called "magic bullet" approach to finding one compound that can specifically treat a dreaded disease.

Crude drugs, consisting of dried plant parts, whole plant extracts, or in some instances isolated compounds, are still used in a few over-the-counter (OTC) drugs. Consumers use OTC drugs for self-diagnosable, self-limiting ailments that a patient can self-medicate with or without a physician's advice. Examples of OTC drugs include psyllium seed, cascara sagrada bark, and senna leaves used in laxative products or pseudo-ephedrine, an alkaloid from *Ephedra* species used as a bronchial dilator in cold and asthma remedies.

In the United States, until recently large pharmaceutical companies have not been interested in developing new plant drugs. According to Varro Tyler, an expert at Purdue University, the reasons for this lack of interest have included the difficulty in obtaining adequate patent protection, difficulty in obtaining raw materials (for example, the problems with taxol described above), problems in putting together research teams from diverse fields, problems in testing and evaluating plant extracts, and researchers' general attitudes toward medicinal plants. As Tyler emphasizes, plant drug development in a strict regulatory environment like that of the United States requires experts in plant taxonomy, ethnobotany, pharmacognosy, biochemistry, analytical chemistry, pharmacology, pharmaceutics, and medicine. Few organizations, including pharmaceutical companies and educational institutions, can bring the expertise of such diverse talent to a single problem.

The primary reason for the lack of emphasis on new-drug research in the United States may be that, with few exceptions, research has not successfully yielded new, marketable drugs. As a result, from the 1960s through the early-1980s most American pharmaceutical companies discontinued research efforts to discover new plant drugs. Instead they encouraged medicinal chemists to model new synthetic compounds in the laboratory.

New Technologies, New Successes

Successful anticancer drugs like taxol, introduced directly or indirectly through the efforts of the NCI's plant screening program, have heightened interest in medicinal plant research in recent years. Major scientific breakthroughs during the 1980s, including enhanced chemical analysis methods and development of automated laboratory technology, brought staggering increases in chemists' abilities to assess new compounds. Prior to these advances, a chemist could isolate and screen about one hundred compounds a year, using several pounds of plant material to carry out each experiment. Now a chemist can do the same kind of experiments with only a few grams of plant material. A single laboratory using micro amounts of material can evaluate thousands of compounds per week in dozens of biological test models. Organizations like the NCI and pharmaceutical companies can screen tens of thousands, rather than hundreds, of plant extracts a year. For example, using existing technologies, from 1958 to 1980 the NCI screened about thirty-five thousand plant species. Using current methods, that many species can be tested in a year. New methods are faster, more accurate, and cheaper.

Advances in biotechnology, such as improved cell-culture methods, allow compounds or plant extracts to be tested against human cancer cells in the laboratory. Twenty years ago scientists conducted screening only on laboratory animals, a procedure that often produced misleading results.

Lobelia

. . . .

Under cultivation, the diminutive woodland herb lobelia produced more biomass and had higher levels of active components than it did in the wild. Lobelia contains a compound, called lobeline, used in over-the-counter drugs to help smokers quit smoking. Photo by Steven Foster.

But dramatic advances in molecular and cellular biology today allow for enhanced approaches to new-drug development, including targeting large numbers of compounds at cell receptor sites in a highly specific manner. This increases the ability to determine the effect of a compound on a specific function or biological activity.

Exploring the potential of plants as a source of compounds for fueling new technology is, however, limited. Medicinal chemists are restricted by the limits of existing computer techniques for applying mathematical modeling to the combination and random synthesis of molecules. Regardless of the potential that diverse new compounds may present, the number that can enter biological screening programs through this method is finite. Mother Nature offers far more chemical diversity than any laboratory can produce. A million medicinal chemists working for a million years to develop synthetic compounds would not produce the chemical diversity that Mother Nature produces in plants.

New-drug development in the United States focuses on single, isolated chemical components (preferably produced synthetically or semisynthetically); this focus increases chances for approval, marketability, patent protection, and profit. Therefore pharmaceutical companies look to plants as sources of single new compounds for drug development, not to the complexity of the natural plant itself.

Medical Alternatives

But interest in medicinal plants does not end with pharmaceutical companies and new-drug development. To gain power over their own health, consumers choose numerous alternative routes, including disease prevention through attention to diet, exercise, and dietary supplements like herbs. Millions of consumers, frustrated at the cost of medical care and the side effects of "wonder drugs," turn to health care alternatives. Although regulations imposed both on product manufacturers and on practitioners limit the demand for herbal medicine and other alternative medical modalities in the United States, there is substantial interest in such medicine. An analysis of alternative, unorthodox, complementary, or unconventional medicine published in the *New England Journal of Medicine* indicated that in 1990 Americans spent $13.7 billion on alternative medicine. In response to that interest, in 1992 the National Institutes of Health established an Office for the Study of Unconventional Medical Practices (now the Office of Alternative Medicine) to evaluate alternative and unconventional medical treatments.

Thousands of herb products are sold in the more than seventy-three hundred health and natural food stores in the United States; they are increasingly found in pharmacies as well. These products, sold as "dietary supplements" or foods, are often used to prevent disease, reduce stress, and treat minor ailments. At least eight hundred medicinal plants are available in stores. Consumer herb product sales in 1991 totalled $653.2 million, up 39.8 percent over the previous year. Although sold as "foods" that do not carry medicinal claims and do not fit neatly into a regulatory category, consumers buy these products for their health benefits. In other industrialized countries, including Germany, France, England, Australia, and Japan, the same or similar products as those that can only be sold as "health foods" in the United States are sold as legitimate drugs.

Genetic Resources and Medicinal Plant Conservation

As worldwide demand for medicinal plants increases, concern for development of sustainable supplies has also increased. A March 1978 conference on medicinal plant conservation addressed problems of severe genetic loss of medicinal plants that extensive destruction of plant-rich habitats (primarily tropical rainforests) causes. This loss is occurring in the face of rapidly increasing global demand for medicinal plants and growing human population. The conference produced the Chiang Mai Declaration, "Saving Lives by Saving Plants," which recognized "the urgent

The flowers of this biennial plant unfold in the evening and on overcast days. Reaching heights of eight feet, evening primrose grows wild in fields and along roadsides throughout the eastern United States, and its edible roots and seeds are collected for use as treatments for atopic eczema and asthma. But indiscriminate collection of wild plants not only puts the species or habitats at risk; it ultimately risks genetic erosion. Photo by Steven Foster.

Evening primrose

need for international cooperation and coordination to establish programmes for conservation of medicinal plants to ensure that adequate quantities are available for future generations."

To determine the dollar value of a single plant species now growing in the United States, researchers at the University of Illinois–Chicago used botanical and prescription survey data along with global studies on plant sources of new drugs to calculate the value. They concluded that the value of any single species is $203 million (1985 dollars). Extrapolating that figure to the estimated 2,067 plant species that could become extinct in the United States by the year 2000, and assuming that one in 125 plant species screened through a thorough pharmacological assessment will become a new drug, the estimated total value of species loss in the United States alone is $3.248 billion (1985 dollars).

Although the potential medicinal value of plant species is often used as a rallying cry for the need to protect endangered species, there has been little research on the medicinal potential of rare and endangered plants in the United States. The problem with wild-harvested medicinal plants in the United States is that there are no published studies showing how many plants can be taken on a "sustainable yield" basis without threatening wild populations. Such information is the basis for regulating hunting of game animals, but there simply are no similar studies for plants. Indiscriminate collection of wild medicinal plants not only puts individual species or habitats at risk, but it ultimately results in genetic erosion.

Issues of biodiversity and loss of tropical rainforest medicinal plants not yet discovered will, and should, remain the focus of international attention. Endangered taxa not involved in commercial trade will remain the focus of American botanists' and plant scientists' conservation efforts. Genetic erosion of medicinal plants already in commercial trade, however, produces a different set of problems. These problems include habitat loss through agricultural or industrial development and overharvest due to high demand or high prices.

Judicious sustainable use of wild plants requires recognizing the diversity inherent in the gene pool of a given species. Genetic material can be selected from this gene pool for multilocation development of crop improvement. Plants can then be selected for desired traits, and agronomic methods can be developed to meet the supply needs of a growing market. To prevent loss of medicinal plant species in the future, and the consequent genetic erosion, medicinal plant conservation ultimately means medicinal plant cultivation.

Medicinal Plant Cultivation

Medicinal plant cultivation equates to medicinal plant conservation because, with few exceptions, most medicinal plants now harvested from wild habitats can be cultivated. For example, in 1994 India proposed that eleven medicinal plant species harvested from wild habitats (primarily in the Himalayas) be restricted in international trade because of the significant declines in wild populations that overharvesting caused. These eleven plants have all been successfully cultivated under suitable commercial conditions.

American ginseng is monitored in international trade under an international treaty known as CITES (Convention on International Trade in Endangered Species of Wild Fauna and Flora). The United States exported over 150,000 pounds of wild-dug ginseng root in 1993. During the same period the country exported more than 1.5 million pounds of cultivated ginseng root, demonstrating clearly that it can be grown successfully in commercial production.

Most goldenseal is harvested from the wild, although it is easily cultivated under conditions similar to the requirements for cultivating ginseng. Ginseng must be propagated by seeds; goldenseal can be propagated by root cuttings. A ginseng crop takes between four and five years to mature; goldenseal can be harvested after three years. A recent decline in wild goldenseal populations has created market shortages and higher prices.

Echinacea

· · · · · · · · · · · · · · · · · · ·

This prairie species, also known as the narrow-leaved purple cone-flower, grows from Texas north through the Dakotas and into nearby Canada. The root is chewed or used in a tea to treat snakebite, spider-bite, flu, and colds. Up to one hundred tons are harvested annually, a rate that may be damaging wild populations. Photo by Steven Foster.

Cultivation assures a consistent, generally predictable supply of plant material without damaging wild populations. One group being damaged by wild harvesting is the prairie species *Echinacea angustifolia.* Up to one hundred tons of the plant are harvested from the wild each year. Commercial supplies of the root come from as many as five species of *Echinacea,* three of which grow only in limited ranges in the Midwest. Farmers successfully cultivate the species, so wild harvest could be regulated or discontinued. But digging of the root continues unabated, and botanists have observed dramatic declines in the species' wild populations.

Cultivation also provides an opportunity to select genetic material from a specific plant group that has high levels of active components, produces large quantities, and enhances the quality of the finished products. A good example is the herb lobelia. Under cultivation this diminutive woodland herb produced more biomass and had higher levels of active components than it did in the wild. Lobelia contains a compound, called lobeline, that was used in OTC drugs to help smokers quit smoking. One study showing the value of cultivated material over wild-harvested plants looked at a poorly drained cultivated field kept clear of weeds for two years. The cultivated plants produced an average lobeline content of 1.05 percent (dry weight), a content about 40 percent higher than the lobeline content of wild plants. The average fresh weight of cultivated plants was 53 grams, but that of wild plants was as low as 1.5 grams; cultivated plants produced up to 129 seed capsules, but wild plants produced between 5 and 25 seed capsules. Cultivated plants averaged 409 seeds per capsule; wild plants had a high of 246 seeds per capsule. Since the highest concentration of lobeline is in the seeds, cultivated plants produced substantially higher yields of active compounds.

The NCI recognizes the need to cultivate plants. To avert supply crises like that experienced with taxol, the agency instituted a policy on raw material procurement for promising experimental anticancer medicinal plants. The program focuses on cultivating plant material at the outset because cultivation of medicinal plants is the key to averting supply problems in product development and to sustaining consistent supplies for international markets.

A History of Medicinal Plants in North America

The flora of North America—encompassing an area north of Mexico that includes the United States, Greenland, and Canada—contains at least 21,757 species of flowering plants. Many of the world's major vegetation formations are represented in this expanse of widely differing climates.

The region's most diverse collection of forest trees is found in the eastern deciduous forest, which covers 11 percent of the continent. This bioregion is where Europeans had the greatest impact during their first two centuries of settlement. A disproportionately high number of commercial American medicinal plants has come from this region. Many of these plants are wide-ranging in the area east of the Mississippi River but are primarily harvested in the southeastern United States, mostly in the southern Appalachian Mountains.

The majority of commercially harvested wild American medicinal plants still comes from the eastern deciduous forest today. Approximately seventy-five plant species from North American forests, particularly forests in the southern Appalachians, trade in international markets. Yet the harvesting of these plants has more to do with cultural and historical patterns than with biological considerations. For example, California, one of the most botanically diverse states, produces relatively few native herbs for trade. Had Europeans settled the West Coast of the United States first instead of the East Coast, most native herbs in commercial trade today would probably come instead from the Pacific states.

Exotic Invaders

Christopher Columbus's arrival in the New World not only ushered in a new era of human migration, it also led to the transoceanic migration of plants through human intervention. Whether brought intentionally or by chance, the addition of exotic plant species to the North American continent has had a dramatic, often negative, impact on native vegetation. The number of exotic species in the northeastern United States and Canada is estimated to range from 20 percent to 30 percent. Weeds, exotic insects, and fungal diseases have damaged North American forests for more than a century, and introduced pathogens or insects have damaged the health of more than 60 percent of the total forested areas of the Northeast.

Weed science focuses largely on eliminating the negative economic effects that invasive exotic plants cause. Some invasive exotics, like catnip, are harvested along roadsides in Virginia and North Carolina for commercial botanical markets. Certain weed species might be controlled in the future by finding a way to exploit them for economic benefit. Kudzu and Japanese honeysuckle are two examples of weeds that might be commercially useful. The root of kudzu, which is used for digestive ailments, and the flowers and stems of Japanese honeysuckle, which are used in cold and flu remedies, are official drugs in the *Pharmacopoeia of the People's Republic of China*. A widely publicized Harvard University study in 1993 catapulted kudzu into the scientific forefront by showing that constituents of the root may offer new choices in the treatment of alcohol abuse. The study confirmed traditional Chinese use of the plant for treatment of patients under the influence of alcohol, highlighting the importance of scientific research on traditional uses of herbal medicines.

Native Groups' Uses of the Forest Pharmacy

Little is known about how Native Americans used the native flora prior to European settlement in North America. Out of necessity, a *materia medica* (encompassing knowledge of the source, preparation, and application of drugs) derived from indigenous plants undoubtedly developed to a relatively high level. Fields and forests were the only pharmacy. Lacking a written language, native groups in North America, like many indigenous peoples, passed information about plant uses through oral traditions. Such oral traditions also largely served to transmit European settlers' uses of American medicinal plants back to continental Europe.

Explorers, missionaries, naturalists, botanists, physicians, government officials, archaeologists, and later, ethnologists all recorded observations

Most current uses of American medicinal plants evolved from native groups' historical uses. In 1895 John W. Harshberger first applied the term "ethnobotany" to the study of "plants used by primitive and aboriginal people." Photo by Steven Foster.

A Hopi herbalist

about native groups' plant uses. Most of these observations were happenstance or casual, and the study of Native American plant uses did not become a distinct academic pursuit until the end of the nineteenth century. In 1895 John W. Harshberger applied the term "ethnobotany" to the study of "plants used by primitive and aboriginal people," and the science of ethnobotany was born. By this time, however, almost all that was to become known about plants used by Native American groups who once inhabited the area of the original thirteen colonies had been recorded, and the tribes themselves exterminated or relocated. Although Native Americans are known to have used 2,147 plant species as medicine, ethnobotanical knowledge about American medicinal plants is fragmented, incomplete, and scarcely a century old.

Native Americans used practically all American medicinal plants that are now traded in world markets or that were formerly listed as official drugs in the *United States Pharmacopoeia*. These include passionflower (used as a sedative), echinacea (used to treat infections), goldenseal (an antibacterial), saw palmetto (used to treat prostate enlargement), mayapple (source of an anticancer drug), black cohosh (used to regulate menopausal symptoms), blue cohosh (used as an aid in childbirth), and many others (see tables 1, 2, and 3). The most famous American medicinal plant is ginseng, which people use to enhance their overall energy level.

Ginseng: Economic Boon, Medical Enigma

American ginseng is, in economic terms, the single most important medicinal plant of the eastern deciduous forest. American ginseng occurs in rich, shaded forests ranging west from Quebec to Manitoba and south to northern Florida, Alabama, and Oklahoma. It is now considered threatened, rare, or locally endangered due to overharvest of the root. European discovery of ginseng in North America was largely serendipitous. In 1709 Father Petrus Jartoux, a Jesuit missionary in northern China who was on a mapping expedition near Korea observed the harvest of *Panax ginseng* root. In 1711 he provided the first authentic westerner's description of Asian ginseng. In 1714 the *Philosophical Transactions of the Royal Society of London* published Jartoux's observations:

> As to the place where this root grows. . . . there is a long tract of mountains, which the thick forests, that cover and encompass them, render almost impassable. On the declivities of these mountains, in the thick forests, on the banks of torrents, or about the roots of trees, and amidst a thousand other different sorts of plants, the ginseng is found. . . . All which makes me believe, that if it is to be found in any other country in the world, it may be particularly in Canada, where the forest and mountains, according to the relation of those that have lived there, very much resemble these here.

In 1715 word of Jartoux's ginseng work reached Father Joseph François Lafitau, a Jesuit missionary who had come to America in 1711 to work among the Mohawks above Montreal. Sparked by Jartoux's claims that ginseng might also be found in Canada, Lafitau discovered American ginseng in 1716. Lafitau sent samples of dried American ginseng roots to Jartoux. By 1720 Jartoux arranged to export American ginseng roots to China, since Chinese medicine had valued ginseng highly for more than two thousand years. The Chinese warmly greeted news of an abundant American supply since wild Chinese ginseng root had become so scarce that it was available only to the emperor and his court.

Export of American ginseng to Asia continues today. In 1993 the United States exported 1,576,269 pounds of cultivated ginseng worth $57,424,000 ($36.43/lb.). The current export market is part of the legacy of Abraham Whisman of Boones Path, Virginia, who began cultivating American ginseng in the 1870s. By 1895 there were about twenty ginseng growers in the country. North American ginseng cultivation is now concentrated in Wisconsin, the Carolinas, Virginia, and British Columbia. Although ginseng is becoming more scarce in the wild, in 1993 the United States exported 153,526 pounds of wild-harvested American ginseng worth $21,770,100

American ginseng

Discovered 275 years ago, American ginseng is one of the best-known American medicinal plants. Since the eighteenth century, much of the wild-harvested and cultivated supplies of American ginseng have been exported to Asian markets. The plant, which is cultivated on a large scale in the Midwest and South, grows in rich woods from Maine to Georgia, west to Oklahoma and Minnesota. Photo by Steven Foster.

($141.80/lb). Over 95 percent of the American ginseng harvest is exported, most of it to Hong Kong. About six pounds of wild-harvested Asian ginseng is dug in northeast China each year, and a single root may sell on the Hong Kong market for $20,000 or more.

First described as a "stimulant" and "tonic," the term "adaptogen" is now used to describe ginseng's action. Norman Farnsworth of the University of Illinois–Chicago defines an adaptogen as:

- A substance that "must be innocuous and cause minimal disorders in the physiological functions of an organism."
- A substance that "must have a nonspecific action," such as the ability of ginseng extracts to modulate stress and improve performance under a wide variety of stressful conditions.
- A substance that "usually has a normalizing action irrespective of the direction of the disease state."

An adaptogen, then, is considered nontoxic and may help improve physical and mental performance in stressed, diseased, or healthy individuals.

Consumers worldwide use ginseng as an adaptogen and stimulant. In the United States people use it as a flavoring for teas and soft drinks, and health food consumers use it as a general tonic. In Chinese medicine, Asian ginseng is considered to have warming properties and American ginseng is said to have cooling properties, so the two are generally used for different purposes. American ginseng is used for its cooling and thirst-quenching effects in summer, and to reduce fevers. Asian ginseng is used generally as a tonic, for its revitalizing properties especially after a long illness. European regulatory authorities permit labeling of Asian ginseng root for medicinal use.

Varro Tyler, in *Herbs of Choice*, uses ginseng as an example to describe the ways in which various countries regulate herb products. Ginseng was an official medicinal in the *United States Pharmacopoeia* until 1882. At that time, trained pharmacists could recommend it as a stimulant and digestive tonic. Today in the United States ginseng, like many herbs, is treated as a food additive for beverage flavoring but it cannot carry claims about its medicinal value. In Germany, Switzerland, and other western European countries, Tyler notes, Asian ginseng is marketed and sold as an over-the-counter drug with adaptogenic effects. In Germany, ginseng root products are labeled for use as a tonic to invigorate people experiencing fatigue, reduced work capacity, and reduced concentration; these products are also used during convalescence. Claims of traditional use are also permitted in Australia and in France where labels may say that ginseng is "traditionally used for transitory weakness." Ginseng is sold as a food in Canada, with the knowledge that it will be used for health purposes. In contrast, the structure of current regulations governing herb products in the United States limits consumers' ability to make choices about their own health because no health claims may be made about these products, even though such claims are accepted in most developed countries.

Evolution of an American Materia Medica in the Colonies

Pre-Revolutionary War observations of medicinal plant use among Native American groups were largely incidental notes scattered among travel accounts, diaries, and floristic works of explorers or naturalists from European academic institutions and museums. Early settlers' medicine, pharmacology, and drugs were largely of European origin. With the rise of a national identity and independence, Americans looked to their native flora for new remedies that could substitute for imported drugs.

Benjamin Smith Barton's late–eighteenth-century *Collections for an Essay towards a Materia Medica of the United States* was the first American survey on native medicinal plants. Barton's work became the basis for development of American medicinal plants in pharmacy and medicine. The seminal *Collections* exemplifies the importance of forest species in the development of American medical botany. Of eighty clearly identified species in the first part of the survey, 80 percent are found exclusively in forest habitats.

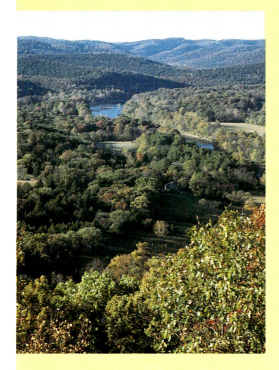

Eastern deciduous forest

· · · · · · · · · · · · · · · · · · · ·

The eastern deciduous forest, a part of which is pictured here, provides the habitat for most of the American medicinal plants that are used in commerce today. Photo by Steven Foster.

Botanist and physician Jacob M. Bigelow wrote *American Medical Botany,* which had more critical analysis of medicinal uses and a more profound impact than Barton's work. Bigelow served on the first United States Pharmacopoeial Committee and produced the list and nomenclature of the first *United States Pharmacopoeia (USP).* The 1820 *USP* was the first official work published under the collective authority of the medical profession. Barton's and Bigelow's works played important roles in introducing indigenous medicinal plants to the first *USP.*

The *USP*'s first revised edition (1830) included 217 official drugs, highlighting the then-growing importance of forest-harvested medicinal plants from the eastern United States. On its primary list the book noted twenty-nine drugs derived from American medicinal plants. Of eighty-seven plants in the secondary list, sixty-four represented American medicinal plants. Most of the native medicinal plants that this early-American pharmacy text listed are still commonly sold as botanicals in the United States today (see tables 1, 2, and 3 for a list of indigenous forest botanicals included in the *USP* and the *National Formulary*).

Mayapple: From Native Remedy to Modern Medicine

One of the most important medicinal plants among Native American groups and still used today is mayapple. Historically the use of most native American medicinal plants, including ginseng and mayapple, resulted initially from observing how Native American groups used the plants. Recognition of the potential medicinal value of mayapple in the eighteenth century helped prompt efforts to discover uses for other American medicinal plants. Mayapple occurs in damp woods from Quebec to Florida and west to Minnesota and eastern Texas. Nineteenth-century medicine used mayapple as a strong laxative to induce vomiting and to expel worms. Mayapple's toxic root treated constipation, jaundice, hepatitis, fevers, and syphilis. Native Americans also used the root to treat cancerous conditions.

The root contains a resin with a toxic compound known as podophyllotoxin. Derivatives of podophyllotoxin are the starting materials for production of semisynthetic drugs used in the treatment of testicular cancer and small-cell lung cancer. Working on the resin obtained from the roots, Jonathan Hartwell, developer and former head of the NCI's Natural Products Branch, discovered an obscure reference to mayapple's potential as a cancer treatment. While he was carrying out a routine literature search on the plant, he discovered that Maine's Penobscot Indians had used it to treat cancer. He also found reference to an 1849 American medical book that recommended use of the resin to treat cancerous tumors and polyps. Hartwell discovered that physicians in Mississippi used it to treat venereal warts (a cancerous condition) as early as 1897, and that Louisianans used it as a folk remedy for treatment of the same condition. Hartwell realized that "both the technical literature and the folklore could be sources of information on use of plants for the treatment of cancer and other growths, and the possibility arose that these records might be useful in providing leads for the laboratory investigation of other plants and the development of useful drugs in the therapy of human cancer."

This realization had a profound though indirect impact on future medicinal plant research in the United States. Colleagues credit Hartwell as being personally responsible for the initiation and early development of the NCI's natural products research program, the program that yielded taxol. He is credited with providing the motivation to create a systematic search for traditional uses of plants and marine animals. However, when it initiated its isolation and screening of potential anticancer compounds from plants in 1961, the NCI chose to develop a random collection program instead of a screening program based on natural products' traditional uses. This random sampling for anticancer activity did not take

American mayapple

. . . .

American mayapple is a common perennial herb found in forests from southern Quebec south to Florida, west to eastern Texas and Minnesota. It has traditionally been used as an anticancer drug; components of the root are used today to produce semisynthetic analogs used in chemotherapy for refractory testicular tumors and in the treatment of small-cell lung cancer. Photo by Steven Foster.

into account plants' historical uses against cancer but instead sought to test plants from botanical families that were known for their toxic compounds.

The genus *Podophyllum,* represented by the North American mayapple and the Himalayan mayapple, presents an example of conservation problems in international trade. Himalayan mayapple is the primary source of mayapple resin, components of which are used to produce semisynthetic cancer treatment drugs. Significant declines of wild Himalayan mayapple populations in 1989, due to root harvest for resin, prompted India to propose the plant for protection under the CITES treaty. The treaty provides that international sale of the dried root must conform to certain trade restrictions, but it does not include chemical components derived from the plant. The listing procedure therefore brought attention to a conservation problem but did little to assure future adequate supplies of the drug or protect wild populations of the plant.

American mayapple is not a significant commercial source of resin. It is easily cultivated and could produce future supplies but it contains lower levels of biologically active constituents than Himalayan mayapple.

Nineteenth-Century Medicine and Forest-Derived Medicinals

John Uri Lloyd, who lived 1849–1936 and is known as the father of the American materia medica, described the development history of American plant-derived drugs as "a struggle for and against authority." Medicine in early–nineteenth-century America included a mix of competing schools of thought that were often in conflict with each another. Orthodox physicians, known as "regular" or "allopathic" medical doctors who were products of the then-emerging medical schools, according to Lloyd used "cherished remedies and methods heired from the past." Another group was composed of European immigrants, calling themselves "Indian doctors" and unorganized as a body, who served as itinerant healers and based treatment on remedies supposedly learned from Native Americans. The Thomsonians, and later botanic physicians, based treatment on popular promotion of limited concepts of medical treatment. "Irregulars" or "liberals" (later called "Eclectics") looked for new-drug development from nontraditional sources, often based on the clinical findings of contemporary peers. Their focus was primarily on developing American medicinal plants. These competing sectarian medical groups created an atmosphere that led to schism and antagonism rather than cooperation.

Lloyd described the antagonism and opposition as "war, for war it was, bitter, vicious, unrelenting, [which] led good men engaged in a humanitarian cause to neglect and oppose kindly methods of medication and led them also to decline to use American drugs, [which] but for professional antagonism might have been broadly introduced at the beginning of the nineteenth century instead of at the close." More than any other group, the Eclectic physicians spurred development of an American pharmacy based on medicinal plants. In the 1880s more than ten thousand people nationwide practiced Eclectic medicine, and Lloyd was the premiere contributor to development of American medicinal plants. He headed Lloyd Brothers Pharmacists, Inc., which manufactured 379 drugs, primarily derived from native plants, that Eclectic practitioners used. Lloyd, the author of more than five thousand articles and eight novels, was among the few individuals who garnered respect from all factions of medical practice. He co-created the Lloyd Library and Museum in Cincinnati, Ohio, which is still the world's largest medicinal plant library.

Goldenseal

One plant that Lloyd promoted was goldenseal. Barton first introduced goldenseal to American medicine, and Lloyd, along with other Eclectic physicians, brought it into broad use. In the first part of his classic *Collections* Barton made one of the earliest observations about cancer among Native Americans: "I am informed that the Cheerake [sic] cure it with a plant which is thought to be the Hydrastis Canadensis, one of our fine native dies [dyes]." In the second part of *Collections* Barton noted that "the Hydrastis is a popular remedy in some parts of the United States." He recorded use of the root as a bitter tonic and as a wash in cold water infusions for eye inflammations. The primary active compounds in goldenseal, the alkaloids berberine and hydrastine, were used until recent years in commercial eye care products.

Goldenseal grows in rich, moist woods favoring beech canopy from Vermont to Minnesota and south to Georgia, Alabama, and Arkansas. It is harvested in large quantities from the eastern deciduous forest for products primarily sold in the United States in health and natural food markets. With current annual usage estimated at about 150,000 pounds of wild-harvested root, there is growing concern over the decline of wild populations.

As early as 1884 Lloyd noted dramatic declines in wild populations as a result of root harvest and habitat loss. However, Lloyd painted a complex picture of the economic and social reasons that helped cause periodic shortages. His observations showed that a decrease in supply did not always or only accompany decreases in habitats or populations. For one thing, farm laborers and the poor historically collected roots during times when they suffered from economic hardship or during years when the crops failed. Not as many people engaged in herb collection during periods of economic prosperity or abundant crops, which would lead to a smaller harvest. And since herbs were a minor commodity, other factors such as increased demand for a new product that would consume the entire supply in one season might also cause shortages and price increases. Shortages in one season stimulated greater harvest the following season, bringing on a subsequent market glut and price decrease. Collectors or dealers were left with overstocks, which they then sold at low cost, providing a disincentive for root harvest during the following season. With prices depressed, collectors turned their attention to other pursuits; once the price stabilized and current stocks were depleted, according to Lloyd, "history repeats itself." Modern shortages are attributed to increased demand coupled with decreasing wild populations caused by overharvesting the root.

Products containing goldenseal have been recommended as tonics, stomachics, antacids, and antispasmodics; they are used for gastritis, colitis, varicose veins, menstrual difficulties, sore mouth or other local irritations, and other ailments. Goldenseal was formerly listed among the official remedies in the *USP* and *National Formulary*. One modern folk use for goldenseal is to mask illicit drugs in urine tests. This use stems from the plot of *Stringtown on the Pike,* a John Uri Lloyd novel, but there is no scientific evidence that hydrastine masks the presence of illicit drugs in urine. In fact, attempts to use goldenseal to mask the presence of morphine have proven to backfire and may instead promote false-positive readings.

Despite its current popularity, goldenseal is little researched. There is no confirmation of its safety and efficacy, or study of the general biology and reproductive needs of the plant itself.

Demise of Sectarian Medical Groups

The attacks that the Thomsonians, Botanics, Eclectics, and other groups made on orthodox medicine, coupled with waning public confidence in the medical profession, gave rise to a counterattack by the medical establishment. This eventually resulted in the demise of all but the orthodox school of medicine. One primary motive behind the formation in 1847 of the American Medical Association (AMA) was the perceived need to lead the public back to orthodox medicine by improving medical education. The AMA recommended that year that physicians receive a minimum of six months' education.

Over the past twenty years, goldenseal has been among the best-selling botanicals in domestic herb markets. It grows in rich, moist, hilly woods from Vermont to Minnesota and south to Georgia, Alabama, and Arkansas. Goldenseal over-harvest is placing pressure on wild populations and stimulating commercial cultivation under artificial shade. Photo by Steven Foster.

Goldenseal

In 1904 the AMA appointed the Council on Medical Education, whose mission it was to upgrade medical colleges. At the time there were 166 medical schools (including orthodox, Eclectic, and other schools). In 1907 AMA representatives visited and classified all schools. Eclectics and other groups complained about their ratings, which led the Carnegie Endowment for the Advancement of Teaching to make an independent and objective rating. In 1909 and 1910 Abraham Flexner for the Carnegie Endowment and Nathan Colwell for the AMA conducted the survey of medical schools. Their published finding was called the "Flexner Report."

The Flexner Report classifications resulted in an ideological direction that determined the future pattern of medical education. One of the report's criteria about curriculum that affected future development of medicinal plant research was that less than 10 percent of a student's time should be devoted to pharmacology and materia medica during the first two years of medical school. State licensing boards gradually accepted the report's findings and ongoing AMA evaluation of medical schools, effectively barring from practice graduates of schools not endorsed by the AMA. This resulted in the demise of Eclectic medicine and along with it the Eclectics' focus on development of American medicinal plants in clinical practice.

Decline of Medicinal Plant Research in the United States, 1930-80

The *USP* first admitted synthetic drugs in 1900. This sparked a new era of drug development and regulation that eventually replaced medicinal plant research in the United States with the study of isolated, synthesized chemical compounds. Coupled with the development of sulfa drugs in the 1930s and penicillin in the early-1940s, the demise of medicinal plant research was the result of serendipitous social, economic, and regulatory factors and not the result of a lack of safety or effectiveness on the part of plant drugs.

A drought of American scientific research on American medicinal plants has persisted at least for the past fifty years. Norman R. Farnsworth and D. D. Soejarto of the University of Illinois–Chicago point out that of five thousand plants species globally that have been scientifically investigated for their potential as drugs, plants in the United States and Mexico are not proportionately represented. They note that although Japanese scientists investigate Japanese plants as drug sources, Russian scientists investigate potential plant drugs from Russia, and so on, "American scientists, for some strange reason, rarely investigate American plants as sources of drugs." Most scientific research on indigenous American species is found in the scientific literature of countries other than the United States, notably Germany and Japan. This is primarily because regulations encourage research in countries that allow herbs to be sold as drugs rather than only as foods.

Regulating Medicinals: Is it Science or Social Reaction?

Forest medicinal plant production in the United States has largely been market driven, following supply and demand trends. Those trends have been influenced by the twentieth-century regulatory environment that has evolved both in the United States and in other countries. In *Herbs of Choice,* Varro Tyler explains that one reason that medicinal plants are not labeled for health purposes on the American market is because of the structure of American drug laws, which require absolute proof of safety and efficacy to bring a "new drug" to market. The cost of developing a single new drug in the United States is estimated at $359 million. To recoup that investment, the government gives pharmaceutical companies exclusive rights to market a new drug for up to twenty-two years. Therefore, plant-drug research in the United States is limited to single chemical compounds (preferably synthesized if the compound is based on a naturally occurring molecule) because single compounds lend themselves better to patent protection.

Even an herb proven safe and effective would provide no incentive to a pharmaceutical company to spend the necessary money on a drug application when another company could then use the research to develop a similar, competing product. And since most herbal drugs have medicinal properties that have long been recognized, they cannot be readily patented. Although chemical entities extracted from plants can be patented, patent laws are complex. Patent requirements revolve around the novelty of the substance, the nature of its usefulness, and experts' views as to its uniqueness.

Still, about eight hundred plant species are found in thousands of products sold through health and natural food stores in the United States. Consumers often purchase these products for the perceived health benefits rather than for flavor or fragrance. Herb product sales for 1991 in the seventy-three hundred health and natural food stores in the United States were estimated at $653.2 million. Including mail order companies, multi-level marketing firms, and other retail sources the U.S. market may be as large as $1.6 billion. The current regulatory system in the U.S. results from the evolution of American drug laws.

Development of U.S. Food and Drug Laws

Responding to unscrupulous industry practices, Harvey W. Wiley, chief of the USDA's Bureau of Chemistry (which later became the Food and Drug Administration [FDA]), rallied against variant industry practices such as mislabeling and adulteration. His efforts resulted in the Food and Drug Act of 1906. This landmark legislation, and its 1912 Sherley Amendment, helped alleviate fraudulent mislabeling and adulteration in the marketplace.

In 1938 Congress passed the Federal Food, Drug, and Cosmetic Act, requiring that drugs be proven safe before entering interstate commerce. Products that had been in commerce before the 1906 Food and Drug Act were "grandfathered" (exempted from the new provisions) and the law did not require additional proof of safety for these products. In 1961 the Kefauver-Harris Amendments to the 1906 act required that drugs be proven both safe *and* effective. This time drugs marketed before 1938 were grandfathered, but products introduced between 1938 and 1962 were subject to the provisions of the new law. Presumably the sale of all American medicinal plants and their respective products were grandfathered since most or all indigenous plants in commerce had been actively marketed prior to 1906.

As a result of the 1961 amendment, the FDA undertook in 1972 a massive review of OTC drug ingredients found in more than three hundred thousand products. OTC drugs are designed to treat self-diagnosable, self-treatable, and self-limiting disease such as headaches or the common cold. The agency convened expert panels to review ingredients in a number of special areas, such as sunscreens, sleep-aids, and a variety of other categories. Under the review process, the agency evaluated each advisory panel's report, along with public comments, and then published a final regulation. The OTC drug review process placed ingredients in one of three categories. Category I meant that the product's ingredients were generally recognized as safe and effective. Category II meant that the ingredients

Mormon tea
· ·
Mormon tea, also known as Brigham tea
or teamster's tea (among other names),
occurs in the desert Southwest. There are
nine species and two hybrids found in
American deserts. Asian *Ephedra* species
are commercial sources of the alkaloids
ephedrine and pseudoephedrine, which are
used for their bronchodilation effects in
over-the-counter cold and allergy formu-
lations. American deserts, which are domi-
nated by woody plants, next to the eastern
deciduous forest are the most important
biome for economically significant
medicinal plants. Photo by Steven Foster.

were not generally recognized as safe and effective, and the product would
be misbranded if labeled as a drug. Category III meant "available data is
insufficient to permit final classification at this time as Category I or II."
Category III products could, however, be marketed if they were sold in
similarly formulated products prior to 1972, if they were considered safe,
if they were not regarded as prescription drugs, and if they did not bear
claims about disease conditions that require professional medical attention (meaning that the product was to be used only for self-limiting
conditions).

Although prescription drug reviews are product specific, the OTC
reviews were ingredient specific. Advisory panels based decisions primarily on data that companies or organizations with an interest in marketing
a particular ingredient supplied. If no one came forward to support an
ingredient, it was either excluded, placed in category II (unsafe or ineffective), or placed in category III (insufficient data to judge effectiveness).
Many good plant drugs were lost in the OTC review process simply because nobody championed the drug. For example, no commercial interest
came forward in favor of retaining prunes as laxatives. As a result, prune
juice cannot be labeled a laxative despite its well-known effect. Of 258
OTC ingredients that the FDA declared ineffective on 16 May 1990, most
were botanical ingredients. The FDA limited the ingredients it reviewed to
those that had been "marketed for a material time and to a material extent

in the United States," which thereby excluded review of well-known medicinal plant preparations available in Europe.

From Drugs to Dietary Supplements

Herb products, which are primarily sold through health and natural food outlets, are positioned in the U.S. market as "dietary supplements." These products are legally considered "foods" rather than "drugs," and product labels do not include information about their medicinal or drug benefits. If manufacturers make medicinal claims, their products become drugs. If a product label provides indications for use, the active ingredient must either be subjected to an OTC monograph or the manufacturer must get approval through the expensive new-drug application process (which as noted above involves an average cost per compound of $359 million). A product in "interstate commerce" that makes a medicinal claim without either of these two approvals would be subject to FDA seizure. This is why herb products, even those that have well-known health benefits, are sold as "foods" and do not make explicit medicinal claims.

Over the past two decades, the FDA has treated herbs and herb products sometimes as "foods" and sometimes as "food additives." Foods are considered safe unless proven otherwise, but food additives, unless they are on the "generally recognized as safe" (GRAS) list, are considered unsafe until the manufacturer proves them to be safe and receives premarket FDA approval. Safe use as a food prior to 1958 probably grandfathers ingredients from the stringent safety requirements of "new" food additives. Herbs that appear on the FDA's GRAS list include familiar herbs such as rosemary, sage, and peppermint, which are often used as natural flavorings. Developed in the late-1950s, the GRAS list reflects scientific consensus about food additives generally recognized as safe at that time. There are about two hundred fifty herbs on the list. Some herbs used in teas, capsulated herb products, tinctures, extracts, and other products are on the GRAS list, but many are not. The FDA generally has not challenged herb products on the GRAS list as long as the label on a product that contains herbs does not make medicinal claims. Advertising or accompanying literature is also considered "labeling" if it is associated with a product. The FDA has attempted to define the term "food additive" broadly, as a way to regulate herbs and herb teas. For example, in 1976 the FDA banned sassafras sold for use in herbal tea on the basis that a cup of hot water is a "food" and sassafras root bark and/or leaves, which are used to flavor the water, are food additives.

Sassafras

. . . .

Nicolas Monardes of Seville in 1574 first detailed the "healing virtues" of sassafras. Sassafras root bark was once used in a popular domestic drink taken as a spring tonic. It was officially listed as a remedy in the *United States Pharmacopoeia* from 1820–1926 and in the *National Formulary* from 1926–65, primarily as a stimulant tonic for digestive problems. Photo by Steven Foster.

The Dietary Supplement Health and Education Act of 1994 lays the foundation to create a rational federal framework for regulating dietary supplements, guaranteeing consumers access to safe and beneficial products and to balanced information about their benefits. Dietary supplements are, for the first time, specifically defined to include vitamins, minerals, herbs or other botanicals, amino acids, or other dietary substances used to supplement the diet by increasing the total dietary intake. Concentrates, metabolites, constituents, extracts, or combinations of these are also included in the definition. Prior to this definition, dietary supplements were defined as "vitamins, minerals, or other ingredients." Herb products ambiguously and vicariously were considered by manufacturers as "other ingredients."

Under the law, dietary supplements are also specifically excluded from being treated as food additives. This means that the FDA can no longer attempt to remove products from the market by claiming that a dietary supplement ingredient has the same legal status as a chemical or preservative added to processed food.

Although existing safety standards in the Food, Drug, and Cosmetic Act are preserved, with additional safeguards added to protect consumers from unreasonable risk of injury, the law clarifies for the first time that the burden of proof that a dietary supplement is adulterated or unsafe falls on the government. In the past, an administrative dictate from the FDA was enough to move against a perceived unsafe product. Now evidence that a dietary supplement is unsafe must be presented before a court for a decision.

Another major provision of the bill for the first time allows timely third-party information, including publications, articles, chapters in books, and scientific literature, to support the sale of dietary supplements. The information cannot be false or misleading. It must not promote a particular manufacturer or product brand. It must present a balanced view of the available scientific information. If the information is displayed in a store, it must be physically separate from the product. It must also be free from any appended information, such as stickers or other information. Literature must be reprinted in its entirety, unless it is the official abstract of a peer-reviewed scientific journal, prepared by the editors or author of the article. The right of retailers to sell books and other publications as part of their business is preserved. Also under this section, the burden of proof for violations falls upon the government.

The law allows product labeling to contain a statement describing how the product's consumption affects "structure or function" in humans or their general well-being. However, it does not allow a manufacturer to make a drug claim for a product. Under this provision it may be possible for a manufacturer to claim that a garlic product helps to reduce cholesterol. However, the manufacturer cannot relate that statement to a disease condition, such as indicating that garlic helps to reduce cholesterol, thereby reducing the risk of heart disease. The product label must also carry the disclaimer: "This statement has not been evaluated by the Food and Drug Administration. This product is not intended to diagnose, treat, cure, or prevent any disease."

If such claims are made on product labels, the manufacturer must have substantiation that the claim is truthful and not misleading, and it must notify the secretary of health and human services within thirty days of making such claim. The phrase "dietary supplement" is required on labels, along with the total quantity of all ingredients.

The legislation mandates the establishment of an Office of Dietary Supplements within the National Institutes of Health to conduct, coordinate, and collect data on dietary supplements and be chief adviser on the subject to the secretary of health and human services. A separate presidential commission will be formed to study and make recommendations on dietary supplement labels, issuing a report within two years. Members of the seven-member panel that will be appointed by the president will include three scientists with "training and experience to evaluate the benefits

Saw palmetto

to health of using dietary supplements and one of such three members shall have experience in pharmacognosy, medical botany, traditional herbal medicine, or other related sciences." The law also states that "members and staff of the Commission shall be without bias on the issue of dietary supplements."

While the new legislation seeks to guarantee availability of products, allow truthful, non-misleading scientific information to be used in conjunction with the sale of the product, and give consumers some information on the product's benefits, it falls short in allowing products to be sold as drugs. Other countries, in contrast, have developed systems for permitting drug claims on medicinal herb products, which may serve as models for future regulation in the United States.

European Experience

European regulatory systems widely recognize phytomedicines, a category of plant-derived drug products. Phytomedicines are therapeutic agents derived from plants or plant parts, including preparations made from them; phytomedicines do not include isolated chemically defined substances. A phytomedicine therefore represents the totality of the medicinal plant or one of its parts (such as the root, leaf, flower, fruit, etc.) rather than a single isolated compound. An herbal preparation is considered an active entity even though it may contain hundreds of chemically defined constituents. Isolated chemical compounds, derived from or extracted from plants, are not considered herbal medicines or phytomedicines in this context. This definition excludes from phytomedicine well-known plant-derived compounds such as menthol (from peppermint) or eucalyptol (from eucalyptus).

Annual retail sales for the European phytomedicine market are estimated at over $6 billion. Fifty percent of those sales are in Germany ($3 billion). One reason that Germany remains the strongest international market for phytomedicines is that a tradition of medicinal plant usage has prevailed there for hundreds of years. One survey revealed that 76 percent of German women drank herbal teas for health benefits, and over 50 percent have turned to herbal remedies for treatment at initial stages of illness.

The German Regulatory Model

Many ingredients that have been deemed ineffective or unsafe in the OTC review process and sold as dietary supplements in the United States carry therapeutic claims when sold in Germany and other western industrialized nations. Complex plant extracts "standardized" to a certain level of active constituents are sold in Germany as phytomedicines. The Germans established an expert commission to develop standardized monographs on herbal medicines. The commission produced nearly three hundred "Therapeutic Monographs on Medicinal Products for Human Use." Each monograph includes details about the drug's name, constituents, indications (including those for the crude drug or preparations), contraindications (if known), side effects (if known), interactions with other drugs or agents (if known), details on dosage or preparations, the method of administration, and the general properties or therapeutic value of the herb or herb product. This system allows manufacturers, within a set of guidelines covering labeling, quality control, and safety, to make well-defined, quality-assured, and properly labeled medicinal herb products publicly available. German manufacturers must supply well-designed pharmacological, toxicological, and clinical data, which for well-known medicinal plants can be supplied through bibliographic information. In such cases, the commission assumes there is a reasonable certainty about safety and efficacy, based on the historical human-use record coupled with documentation in modern scientific literature, including practitioners' clinical experience.

Germany's tolerant regulatory environment for phytomedicines is supported by medical training, since medical students in Germany take a

Bloodroot

specified number of course hours in phytomedicine and are tested on the subject in their licensing exam. More than 70 percent of general practitioners in Germany prescribe phytomedicines. The German monograph system is the best governmental information source on western medicinal plant usage. There are currently about sixty thousand phytomedicine products on the German market, and this well-developed regulatory system serves as a primary model for regulation of medicinal plant products throughout the European Union.

The Canadian Model

A positive regulatory environment for herb products is not unique to Germany or other European countries. Manufacturers in Canada can apply for a drug identification number (DIN), which is a regulatory category equivalent to the OTC category in the United States. If literature reviews, recent well-designed clinical trials, and other necessary data establish the product's safety and efficacy, a DIN can be awarded and specific medicinal claims placed on the herb product's label. For example, a product made from the leaves of feverfew may carry claims that it alleviates symptoms of migraine headaches. Based on traditional usage in England for the treatment of migraine headaches, several clinical studies confirmed that at specified dosage levels the herb reduced the duration and severity of migraines while causing only minor, reversible side effects. An advisory committee has recommended that Canada also establish a "traditional medicines" category, which would permit traditional claims about a product if those claims are well-documented and generally accepted. Such traditional-use claims are permitted in England and France.

World Health Organization Initiatives

The World Health Organization (WHO) in 1978 instituted a comprehensive approach to medicinal plant development worldwide. WHO advocated the development of an inventory and classification system for medicinal plants used in different countries. It called for adopting scientific criteria and methods for assessing the safety and efficacy of medicinal plant products in disease treatment and prevention. WHO also designated collaborative research and training centers throughout the world and developed programs for disseminating information. WHO's overriding concern has been to insure that herbal medicines are safe, in whatever form they are made available.

In 1987 the Fortieth World Health Assembly instituted new action mandates, including these goals:

• To initiate comprehensive programs for the identification, evaluation, preparation, cultivation and conservation of medicinal plants used in traditional medicine.
• To ensure quality control of drugs derived from traditional plant remedies by using modern techniques and applying suitable standards and good manufacturing practices.

WHO published *Guidelines for the Assessment of Herbal Medicines* in 1991 to address several major areas regarding assessment, safety, and quality assurance. The areas addressed included:

• Pharmaceutical assessments to suggest ways to document authentication and quality control of crude plant material, plant preparations, control of finished products, and product stability.
• Recommendations for safety assessments, including the significance of toxicological studies and the value of safety documentation based on experience.
• Suggestions for the assessment of efficacy and intended use, documentation of pharmacological activity, and evidence required to support medicinal use of herbs.

Passionflower is more widely respected in Europe than in its native North America. Europeans use the fresh or dried whole plant to treat nervous tension and anxiety, including cases of sleep disturbance. The plant, a perennial vine with an extraordinary flower, grows from Virginia west to southern Illinois and southeast Kansas, south to Florida and Texas. Photo by Steven Foster.

Passionflower

The *Guidelines* represent the first official document to provide comprehensive and uniform model guidelines for the benefit of national health authorities, industry, academic communities, and others concerned with the availability, quality, and use of herbal medicines. WHO has been a driving force in stimulating collaborative research in traditional medicine with new scientific research in order to bring health care to all people by the year 2000. It has been responsible for motivating and implementing national surveys of medicinal plants and the evaluation of those plants, and it has developed guidelines for international initiatives on medicinal plant conservation, utilization, regulation, and integration into public health policy.

Challenges for the Future

There is a worldwide renaissance in medicinal plant research, sparked by successes such as implementation of WHO's Traditional Medicine Programme, which has helped empower traditional healing systems in developing countries. The success of new drugs like taxol, along with technological advancements in molecular biology, biotechnology, and laboratory automation, have renewed pharmaceutical industry interest in plants as sources of compounds for new drugs. Regulatory acceptance of phytomedicines in developed countries, including Germany and Japan, has helped stimulate independent research on active principles, mechanisms of action, safety, and clinical effectiveness of medicinal plants.

There is no regulatory mechanism for bringing complex plant extracts, which have little chance of being patented, to the American marketplace. In contrast, Germany, Japan, Canada, and other developed countries take a more lenient approach to regulating whole plant parts, their extracts, or their derivative products. These countries consider the product's history of safe use along with documentation of its clinical efficacy.

Approximately seventy-five indigenous American medicinal plants enter international commerce. Most come from forest habitats. In countries such as England, France, and Germany, these American medicinal plant products carry therapeutic claims. In the United States they are called dietary supplements and by law can carry no substantive label information to help consumers make informed choices about their use. A U.S. regulatory mechanism based on something like the German model would instead recognize a product's established safety and efficacy, and assure its quality by requiring production standards of identification and purity. The American public, segments of the scientific community, and manufacturers would all benefit from such a system.

A healthy future for medicinal plants requires establishing data to determine sustainable yields of wild populations and developing commercially cultivated supplies. A 1993 Harvard study published in *The New England Journal of Medicine* on unconventional medicine in the United States, measuring its prevalence, costs, and patterns of use, indicates that the public demands and seeks medical alternatives for disease prevention and treatment. The rapid growth of herb sales in health and natural food markets also indicates strong consumer demand.

The history of the American forest pharmacy provides many examples of successful clinical use of native medicinal plants to prevent and treat disease. That historical record can be tapped in future development of drugs or simply to develop new ways of using botanicals already available in the marketplace. Some people envision for the future a regulatory mechanism that assures proper identity of herbs used in products, permits labeling that includes substantiated health claims, and requires that herb products meet established quality and purity standards.

Varro Tyler, in "The Herbal Regulatory Dilemma: A Proposed Solution," presents one logical approach to herb product regulation:

• Prepare a Botanical Codex or similar compendium to establish standards of identity, purity, and quality for all crude vegetable drugs
• Identify herbs on packaging by their Latin binomial

- Institute a system to determine that manufacturers comply with appropriate quality and purity standards
- Establish safety guidelines for herbs sold to consumers
- Permit sales of herbs that have approved traditional claims of efficacy, provided other requirements have been met

A regulatory mechanism encompassing the above points would provide truthful information on product labels and contribute to public health care in the United States. It would also help place medicinal plants more squarely within mainstream medicine, where they have long been the progenitors of effective treatment.

TABLE 1 Trees listed for their medicinal value in the *United States Pharmacopoeia* and the *National Formulary*

LATIN NAME	COMMON NAME	FORM	MEDICINAL VALUE	USES
Abies balsamea (L.) Mill.	Canada balsam fir	Oleoresin	Antiseptic	To treat hemorrhoids; also used as a root canal sealer
Betula lenta L.	Birch	Oil of sweet birch	Anti-inflammatory, analgesic	Once used to treat rheumatism, gout, scrofula, bladder infection, neuralgia, **toxic**
Castanea dentata (Marsh.) Borkh.	Chestnut	Leaves	Astringent	Once used to treat whooping cough; also used as a placebo for children
Chionanthus virginicus L.	Fringe-tree	Bark of root	Diuretic, alterative, cholagogue	Once used to treat congestion of glandular organs and the venous system, particularly liver disease
Cornus florida L.	Dogwood	Bark	Astringent	Once used in the South, especially during the Civil War, to treat malarial fevers (substitute for cinchona, source of quinine); also to treat chronic diarrhea
Diospyros virginiana L.	Persimmon	Fruit	Astringent	Once used to treat diarrhea, dysentery, uterine hemorrhage
Fagus grandifolia Ehrh.	Creosote, beechwood	Creosote	Irritant, styptic, antiseptic	Once used to treat eruptions, wounds, ulcers, **toxic**
Fraxinus americana L.	White ash	Bark	Emetic, laxative	Once used to treat menstrual problems
Juglans cinerea L.	Butternut	Bark of root	Laxative	To treat constipation
Juniperus virginiana L.	Red cedar	Wood	Diaphoretic	Once used to treat colds, coughs; also used to induce sweating
Liquidambar styraciflua L.	Styrax	Resin	Antiseptic, expectorant, antimicrobial, anti-inflammatory	Used as a topical antiseptic

TABLE 1 Trees

LATIN NAME	COMMON NAME	FORM	MEDICINAL VALUE	USES
Liriodendron tulipifera L.	Tulip-tree	Bark	Astringent, antiperiodic	Once used to treat malarial fevers (substitute for cinchona, source of quinine); also to treat coughs, dysentery, rheumatism
Magnolia virginiana L., *Magnolia glauca* L.	Sweet bay	Bark	Astringent, antiperiodic	Once used to treat malarial and typhoid fevers (substitute for cinchona, source of quinine); also to treat indigestion, rheumatism, worms
Pinus palustris Mill.	Pine tar	Oil of tar, rectified	Antipruritic	Once used to treat skin diseases such as eczema, psoriasis, **toxic**
	Turpentine	Oil of turpentine	Rubefacient	Once used to treat gangrene, frostbite, burns, carbuncles, ulcers, bronchial catarrh, inflamed urinary passages, **toxic**
	Long leaf pine	Tar	Antipruritic	Once used to treat eczema, psoriasis, scabies, **toxic**
	Turpentine	Turpentine	Stimulant, counterirritant	Once used to treat rheumatism, **toxic**
Pinus strobus L.	White pine	Bark	Antitussive	Once used to treat coughs
Populus balsamifera L., *Populus tacamahaca* Miller	Balm gilead	Buds	Anti-inflammatory	Once used to treat toothaches, rheumatism, diarrhea, wounds, coughs, pleurisy
Populus tremuloides Michx. & spp.	Wood charcoal	Wood charcoal	Absorbent	To treat indolent ulcers and absorb poisons in stomach; also used as a dentifrice

table 1 continues

TABLE 1 **Trees**

LATIN NAME	COMMON NAME	FORM	MEDICINAL VALUE	USES
Populus tremuloides Michx. & spp.	Salicin	Salicin	Anti-inflammatory, analgesic	Once used to treat malarial fevers (substitute for cinchona, source of quinine); also to treat acute rheumatism
Prunus serotina Ehrh.	Wild cherry	Bark	Expectorant	Once used to treat coughs, fevers, colds, sore throats, diarrhea, bronchitis, pneumonia, dyspepsia
Quercus alba L.	White oak	Bark	Astringent	Once used to treat diarrhea, dysentery
Quercus velutina Lam., *Quercus tinctoria* Bartr.	Black oak	Bark	Astringent	Once used to treat diarrhea, dysentery
Rhamnus purshiana DC.	Cascara sagrada	Aged bark	Laxative	To treat constipation
Sassafras albidum (Nutt.) Nees, *Sassafras variifolium* (Salisb.) Ktze. *Sassafras officinale* Nees & Eberm.	Sassafras	Bark of root	Aromatic	Once used to treat stomachaches, indigestion, fevers
		Pith	Demulcent	Once used to treat sore throats, mucous membrane irritation, eye ailments
Thuja occidentalis L.	Arbor vitae	Bark	Astringent	Once used to treat worms, rheumatism
	Cedar	Oil of cedar leaf	Antiseptic, expectorant, counterirritant	Once used to treat rheumatism, **toxic**
Ulmus rubra Muhl., *Ulmus fulva* Michx.	Slippery elm	Bark	Demulcent, nutritive	To treat sore throats, bloody diarrhea, ulcers
Zanthoxylum clava-herculis L.	Prickly-ash	Berries	Diuretic	Once used to treat sore throats, tonsillitis

TABLE 2 Herbaceous plants listed for their medicinal value in the United States Pharmacopoeia and the National Formulary

LATIN NAME	COMMON NAME	FORM	MEDICINAL VALUE	USES
Aletris farinosa L.	Aletris	Root	Anti-inflammatory, tonic	Once used to treat dyspepsia, diarrhea, rheumatism, jaundice
Aralia nudicaulis L.	American sarsaparilla	Root	Stimulant, diaphoretic	Once used as a "spring tonic"; also used as a blood purifier
Aralia racemosa L.	Spikenard	Root	Stimulant, expectorant	Once used to treat coughs, asthma, irritation of the brochopulmonary tract
Arisaema atrorubens (Ait.) Blume, *Arisaema triphyllum* (L.) Schott.	Arum, wild turnip	Root	Expectorant, diaphoretic, purgative	Once used to treat asthma, bronchitis, colds
Aristolochia reticulata Nutt., *Aristolochia serpentaria* L.	Serpentaria	Root	Stimulant, tonic, diaphoretic, diuretic	Once used to treat fevers, dyspepsia, **toxic**
Asarum canadense L.	Canada snakeroot	Root	Aromatic, stimulant, diaphoretic	Once used to treat indigestion, fevers, colds, sore throats, menstrual irregularities
Caulophyllum thalictroides (L.) Michx.	Blue cohosh	Root	Emmenagogue	Once used to treat profuse menstrual flow, abdominal cramps, urinary tract infections; also used as an aid in labor
Chamaelirium luteum (L.) Gray	Helonias	Root	Stimulant, diaphoretic	Once used to treat indigestion, menstrual irregularities
Chimaphila umbellata (L.) Nutt., *Chimaphila maculata* (L.) Pursh.	Pipsissewa	Herb	Astringent, diuretic	Once used to treat urinary tract infections
Cimicifuga racemosa (L.) Nutt.	Black cohosh	Root	Estrogenic, sedative, anti-inflammatory	To treat menstrual irregularities

table 2 continues

TABLE 2 Herbaceous plants

LATIN NAME	COMMON NAME	FORM	MEDICINAL VALUE	USES
Collinsonia canadensis L.	Stone root	Root	Diuretic, astringent	Once used to treat indigestion, diarrhea, kidney/bladder ailments, laryngitis, menstrual disorders
Coptis groenlandica (Oeder) Fernald	Goldthread	Root	Astringent, anti-inflammatory	Once used to treat canker sores, infections
Cypripedium calceolus L. var. *pubescens* (Willd.) Correll, *Cypripedium parviflorum* Salisb. or *Cypripedium parviflorum pubescens* (Willd.) Knight	Lady-slipper	Root	Sedative	Once used to treat menstrual irregularities
Delphinium canadensis (Goldie) Walp	Corydalis	Herb	Antispasmodic, diuretic	Once used to treat asthma, edema, **toxic**
Dicentra cucullaria (L.) Bernh.	Corydalis	Herb	Diuretic, diaphoretic	To treat syphilitic conditions
Dioscorea villosa L.	Wild yam	Root	Antispasmodic	Once used to treat colic, gas, rheumatism
Drosera anglica Huds., *Drosera longifolia* L., *Drosera rotundifolia* L.	Sundew	Herb	Antitussive	To treat coughs, bronchitis, asthma
Dryopteris filixmas (L.) Schott, *Dryopteris marginalis* (L.) A. Gray	Aspidium (male fern)	Root, oleoresin	Vermifuge	Once used to treat intestinal parasites, **toxic**
Erigeron philadelphicus L.	Daisy fleabane	Herb	Astringent, diuretic	Once used to treat diarrhea, gravel, diabetes
Erythronium americanum Ker.	Yellow adder's-tongue	Herb	Emetic, emollient	Once used to induce vomiting
Eupatorium perfoliatum L.	Boneset	Herb	Diaphoretic, astringent	Once used to treat fevers, flu epidemics

TABLE 2 **Herbaceous plants**

LATIN NAME	COMMON NAME	FORM	MEDICINAL VALUE	USES
Eupatorium purpureum L.	Gravel root	Root	Astringent, diuretic	Once used to treat gravel
Eupatorium verbenaefolium Michx., *Eupatorium teucrifolium* Willd.	Wild horehound	Herb	Diaphoretic, diuretic, laxative, stimulant	Once used to treat fevers
Euphorbia corollata L.	Flowering spurge	Root	Emetic, diaphoretic, expectorant	To treat ascites, diarrhea, chronic bronchitis, **toxic**
Euphorbia ipecacuanha L.	Ipecac spurge	Root	Emetic	To induce vomiting
Euphorbia pilulifera L.	Euphorbia	Herb	Antiasthmatic	Once used to treat chronic bronchitis, cardiac diseases, **toxic**
Frasera caroliniensis Walt.	American colombo root	Root	Emetic, purgative, stimulant	Once used to treat circulatory diseases
Gentiana saponaria L., *Gentiana catesbaei* Walt.	Blue gentian	Root	Digestive	To treat digestive problems
Geranium maculatum L.	Cranesbill	Root	Astringent, styptic	Once used to treat dysentery, canker sores, ulcers
Geum rivale L.	Water avens	Root	Astringent	Once used to treat diarrhea, hemor-rhages
Gillenia stipulata (Muhl.) Trel., *Gillenia stipulacea* Nutt., *Gillenia trifoliata* (L.) Moench.	American ipecac, Indian physic	Root	Emetic, cathartic	Once used to treat constipation, worms, dyspepsia
Hedeoma pulegioides (L.) Pers.	Pennyroyal	Herb, oil of Penny-royal	Emmenagogue	Once used to treat menstrual irregu-larities, **toxic**
Hepatica acutiloba DC., *Hepatica americana* (DC.) Ker., *Hepatica tribola* Chaix.	Liverwort	Leaves	Astringent	Once used to treat liver ailments, fevers, coughs

table 2 continues

TABLE 2 Herbaceous plants

LATIN NAME	COMMON NAME	FORM	MEDICINAL VALUE	USES
Heracleum lanatum L.	Cow parsnip	Root	Antispasmodic, carminative	Once used to treat dyspepsia, flatulence, colic, asthma
Heuchera americana L.	Alum root, sanicle	Root	Astringent, styptic	Once used to treat hemorrhage, dysentery, canker sores, ulcers
Hydrastis canadensis L.	Goldenseal	Root	Astringent, tonic, antiseptic	To treat inflammation, liver disease, infections
	Hydrastine	Salt	Diaphoretic, antiseptic, astringent, styptic	To treat malaria, eye inflammations, uterine hemorrhage
Iris versicolor L., *Iris virginica* L., *Iris caroliniana* Wats	Blue flag	Root	Alterative, cholagogue, stimulant	Once used to treat diseases of the gastrointestinal canal and the glandular and nervous systems, **toxic**
Lactuca virosa L., *Lactuca scariola* L. var. *integrata* Gren. & Godr.	Lactucarium, lettuce opium	Leaf/stem exudate	Sedative, calmative, hypnotic	Once used to treat nervous conditions
	Wild lettuce	Leaves	Sedative, calmative, hypnotic	Once used to treat nervous conditions
Lobelia inflata L.	Lobelia	Herb	Emetic, diaphoretic, antispasmodic	To treat fevers, asthma, **toxic**
Lycopodium clavatum L.	Lycopodium	Spores	Protective	Once used to treat erysipelas, impetigo, herpes, ulcers, eczema
Lycopus virginicus L.	Bugleweed	Herb	Sedative	Once used to treat cardiac disease, insomnia, chronic lung disease
Menyanthes trifoliata L.	Buckbean	Leaves	Astringent, cathartic	Once used to treat dyspepsia, rheumatism, worms, malaria, **toxic**
Mitchella repens L.	Squaw vine	Herb	Astringent, diuretic	Once used to treat edema, diarrhea, urinary difficulties

TABLE 2 Herbaceous plants

LATIN NAME	COMMON NAME	FORM	MEDICINAL VALUE	USES
Panax quinque-folius L.	Ginseng	Root	Tonic, stomachic	To treat general debility, to speed convalescence; also used to improve performance
Passiflora incarnata L.	Passionflower	Herb	Sedative	To treat nervous conditions, insomnia
Phytolacca americana L., *Phytolacca decandra* L.	Poke	Berries, root	Emetic, cathartic	Once used to treat rheumatism, tuberculosis, **toxic**
Podophyllum peltatum L.	Mayapple	Root	Cathartic, stimulant	Once used to treat chronic hepatitis, **toxic**
	Podophyllin	Resin	Cathartic	To treat testicular and small-cell lung cancer, **toxic**
Polygala polygama Walt., *Polygala rubella* Willd., *Polygala senega* L.	Bitter polygala, senega	Root	Emetic, expectorant, cathartic, diuretic, antispasmodic	Once used to treat menstrual irregularities, colds, coughs, convulsions, rheumatism, asthma, pneumonia
Polypodium vulgare L.	Polypodium	Root	Pectoral, demulcent, purgative, anthelmintic	Once used to treat lung disease, liver disease, worms, **toxic**
Sanguinaria canadensis L.	Bloodroot	Root	Emetic, stimulant	Once used to treat chronic hepatitis, atonic dyspepsia, **toxic**
Scutellaria lateriflora L.	Scullcap or Skullcap	Herb	Sedative, nervine, antispasmodic	To treat nervous conditions, insomnia
Senecio aureus L.	Liferoot, groundsel	Root	Emmenagogue, diuretic, diaphoretic	Once used to treat menstrual irregularities, fevers, **toxic**
Solanum carolinense L.	Horsenettle	Berries	Antispasmodic	Once used to treat tetanus, convulsions, epilepsy, **toxic**
Solidago odora Ait.	Goldenrod	Herb	Aromatic, stimulant, diaphoretic, diuretic	Once used to treat diarrhea, fevers, dyspepsia, kidney disease, bladder disease

table 2 continues

TABLE 2 Herbaceous plants

LATIN NAME	COMMON NAME	FORM	MEDICINAL VALUE	USES
Spigelia mari-landica L.	Pink root	Root	Anthelmintic	Once used to treat worms, **toxic**
Stillingia sylvatica L.	Stillingia	Root	Emetic, cathartic	Once used to treat liver disease, syphilis, bronchitis
Swertia chirayita (Roxb.) Lyons	Chireta	Herb	Tonic	Once used to treat urinary complaints, digestive disorders
Symplocarpus foetidus (L.) Nutt, *Dracontium foetidum* L.	Skunk cabbage	Root	Antispasmodic, narcotic	Once used to treat asthma, whooping cough, nervous irritability, **toxic**
Trillium erectum L.	Beth root	Root	Astringent, antiseptic, emmenagogue	Once used to treat menstrual irregu-larities, bronchial afflictions; also used as an aid in labor
Triosteum perfoliatum L.	Fever root	Root	Diaphoretic, emetic	To treat fevers, inflammations, rheumatism
Turnera aphro-disiaca Ward, *Turnera diffusa* Willd. & Schul.	Damiana	Herb	Aphrodisiac	Once used to treat impotence, genito-urinary disease
Urtica dioica L.	Nettle	Herb	Astringent, diuretic	To treat diarrhea, hemorrhoids, fevers, gravel
Veratrum viride Ait.	American hellebore	Herb	Sedative	Once used to treat heart disease, **toxic**
Verbena hastata L.	Vervain	Herb	Expectorant, febrifuge	Once used to treat fevers, menstrual irregularities, gravel
Veronicastrum virginicum (L.) Farw., *Leptandra virginica* L.	Leptandra, culver's root	Root	Laxative	Once used to treat atonic states of the colon
Viola pedata L.	Violet	Herb	Laxative, diuretic, expectorant	To treat uric acid gravel, eczema, asthma

TABLE 3 Shrubs listed for their medicinal value in the *United States Pharmacopoeia* and the *National Formulary*

LATIN NAME	COMMON NAME	FORM	MEDICINAL VALUE	USES
Aralia spinosa L.	Angelica-tree	Bark	Analgesic, counter-irritant	Once used to treat toothaches, rheumatism
Arctostaphylos uva-ursi (L.) Spreng.	Uva ursi	Leaves	Diuretic, astringent	To treat cystitis, nephritis, urethritis, gravel, kidney stones, gall stones
Cornus amomum Mill., *Cornus sericea* L., *Cornus rugosa* Lam.	Swamp dogwood	Bark	Astringent, tonic, stimulant	Once used in the South, especially during the Civil War, to treat malarial fevers (substitute for cinchona, source of quinine); also to treat chronic diarrhea
Eriodictyon californicum (H. & A.) Greene	Yerba santa	Leaves	Expectorant, antispasmodic	Once used to treat asthma, chronic bronchitis
Euonymus atropurpureus Jacq.	Wahoo	Bark of root	Laxative, diuretic, expectorant	To treat fevers, stomachaches, constipation, lung ailments, liver congestion; also used in heart medications, **toxic**
Gaultheria procumbens L.	Wintergreen	Leaves	Astringent	Once used to treat colds, headaches, stomachaches, fevers, kidney ailments, rheumatism, sore muscles
		Oil of gaultheria	Analgesic, carminative, anti-inflammatory, antiseptic	Once used to treat rheumatism, **toxic**
Gelsemium sempervirens (L.) Ait. f.	Gelsemium	Herb	Central nervous system depressant, analgesic, antispasmodic	Once used to treat inflammatory conditions of the cerebrospinal column, **toxic**
Hamamelis virginiana L.	Witchhazel	Bark	Astringent	To treat bruises, sore muscles, hemorrhoids, profuse menstrual flow

table 3 continues

TABLE 3 **Shrubs**

LATIN NAME	COMMON NAME	FORM	MEDICINAL VALUE	USES
Hamamelis virginiana L.	Witchhazel	Leaves	Astringent	To treat hemorrhoids, irritations, minor pain, itching
Hydrangea arborescens L.	Hydrangea	Root	Diuretic	Once used to treat kidney stones, mucous irritation of bladder, bronchial afflictions
Ilex verticillata (L.) Gray	Black alder	Bark	Astringent	Once used to treat fevers
Juniperus communis L.	Juniper	Berries	Diuretic, antiseptic	Once used to treat urinary problems, cystitis, intestinal infections, coughs, stomachaches, colds, bronchitis
		Oil of juniper	Diuretic, antiseptic	Once used to treat cystitis, **toxic**
Mahonia aquifolium (Pursh) Nutt.	Berberis	Root	Astringent, antiseptic	Once used to treat jaundice, hepatitis, fevers, hemorrhage, diarrhea, arthritis, rheumatism, sciatica
Mahonia nervosa (Pursh) Nutt.	Oregon grape	Root	Astringent, antiseptic	Once used to treat jaundice, hepatitis, fevers, hemorrhage, diarrhea, arthritis, rheumatism, sciatica
Menispermum canadense L.	Yellow parilla	Root	Laxative, diuretic	Once used to treat indigestion, rheumatism, arthritis, syphilis, general debility, chronic skin infections, **toxic**
Myrica cerifera L., *Myrica pennsyl - vanica* Liosel., *M. caroliniensis* Mill.	Bayberry	Bark of root	Astringent, emetic	Once used to treat chronic gastritis, diarrhea, dysentery, abnormal vaginal discharges, jaundice, indolent ulcers
Rhus glabra L.	Sumach	Berries	Astringent, antiseptic	Once used to treat diarrhea, dysentery, fevers, mouth ulcers, throat ulcers, leukorrhea, anal prolapse, uterine prolapse

TABLE 3 **Shrubs**

LATIN NAME	COMMON NAME	FORM	MEDICINAL VALUE	USES
Rubus allegheniensis Porter, *Rubus canadensis* L., *Rubus cuneifolius* Pursh	Blackberry	Bark of root	Astringent	Once used to treat diarrhea, dysentery, stomach pain, gonorrhea, back pain
Rubus allegheniensis Porter, *Rubus nigrobaccus* Bailey	Blackberry	Berries	Astringent, flavoring	To treat stomach-aches
Rubus trivalis Michx.	Dewberry	Bark of root	Astringent	Once used to treat diarrhea, dysentery, stomach pain, gonorrhea, back pain
Rubus villosus Ait., *Rubus flagellaris* Willd.	Blackberry	Bark of root	Astringent	Once used to treat diarrhea, dysentery, stomach pain, gonorrhea, back pain
Sambucus canadensis L.	Elder	Flowers	Mild stimulant, carminative, diaphoretic	Once used to treat colds, fevers, flu
Serenoa repens (Bartr.) Small, (*Serenoa serrulata* [Michx.]) Hook. f.	Saw palmetto	Berries	Anti-inflammatory	To treat prostate enlargement or inflammation, colds, coughs, asthma, chronic bronchitis, migraines
Spiraea tomentosa L.	Hardhack	Root	Astringent, analgesic	To treat fevers, minor pain, diarrhea, dysentery
Toxicodendron radicans Mill, *Rhus toxicodendron* Auth. not L.	Poison ivy, poison oak	Leaves	None	Once used to treat paralytic and liver disorders, **toxic**
Viburnum prunifolium L., *Viburnum rufidulum* Raf.	Blackhaw	Bark of root, stem	Uterine tonic, sedative, antispasmodic	Once used to treat painful menses, spasms after labor; also to prevent miscarriage

table 3 continues

TABLE 3 **Shrubs**

LATIN NAME	COMMON NAME	FORM	MEDICINAL VALUE	USES
Viburnum opulus L. var. *americanum* (Mill.), *Viburnum trilobum* Marshall	Cramp bark	Bark	Astringent, anti-spasmodic, sedative	Once used to treat painful menses
Xanthorhiza simplicissima Marsh., *Zanthorhiza apiifolia* L'Her.	Yellow root	Root	Anti-inflammatory, astringent, hemostatic, antimicrobial, anticonvulsant	Once used to treat stomach ulcers, colds, jaundice, cramps, sore mouth or throat, menstrual irregularities
Zanthoxylum americanum Mill.	Prickly ash	Bark	Analgesic, diuretic, astringent	Once used to treat chronic rheumatism, dyspepsia, dysentery, kidney trouble, colds, coughs, lung ailments, nervous debility

Table 4 **Latin names for plants listed in text by common name**

American ginseng	*Panax quinquefolius*
Asian ginseng	*Panax ginseng*
Belladonna	*Atropa belladonna*
Black cohosh	*Cimicifuga racemosa*
Blue cohosh	*Caulophyllum thalictroides*
Cascara sagrada	*Rhamnus purshiana*
Catnip	*Nepeta cataria*
Echinacea	*Echinacea augustifolia*
English yew	*Taxus baccata*
Eucalyptus	*Eucalyptus*
Feverfew	*Tanacetum parthenium*
Foxglove	*Digitalis* species
Garlic	*Allium sativum*
Goldenseal	*Hydrastis canadensis*
Himalayan mayapple	*Podophyllum hexandrum*
Japanese honeysuckle	*Lonicera japonica*
Japanese yew	*Taxus cuspidata*
Kudzu	*Pueraria lobata*
Lobelia	*Lobelia inflata*
Madagascar periwinkle	*Catharanthus roseus*
Mayapple	*Podophyllum peltatum*
Opium poppy	*Papaver somniferum*
Pacific yew	*Taxus brevifolia*
Passionflower	*Passiflora incarnata*
Peppermint	*Mentha piperita*
Rosemary	*Rosmarinus officinalis*
Sage	*Salvia offiinalis*
Sassafras	*Sassafras albidum*
Saw palmetto	*Serenoa repens*
Senna	*Cassia marilandica*

Further Reading

Akerele, O. "WHO Guidelines for the Assessment of Herbal Medicines." *HerbalGram* 28 (1993): 13–20.

Akerele, O., V. Heywood, and H. Synge, eds. *Conservation of Medicinal Plants.* Proceedings of the Chiang Mai Consultation, Chiang Mai, Thailand, March 1978. Cambridge, England: Cambridge University Press, 1991.

Barton, Benjamin Smith. *Collections for an Essay Towards A Materia Medica of the United States.* Bulletin of the Lloyd Library, No. 1, Reproduction Series, No. 1. 1798 and 1804. Reprint (2 vols. in 1), Cincinnati, Ohio: Lloyd Library, 1900.

Beckstrom-Sternberg, S. M. and J. A. Duke. "Ethnopharmacology of Yew." *International Yew Resources Conference. Yew (Taxus) Conservation Biology and Interactions.* Edited by C. R. Temple. Berkeley, California: Native Yew Conservation Council, 1993, p. 3.

Berman, A. "The Impact of the Nineteenth-Century Botanical-Medical Movement on American Pharmacy and Medicine." Ph.D. diss., University of Wisconsin, Madison, 1954.

Bigelow, J. *American Medical Botany.* 3 vols. Boston, Massachusetts: Cummings and Hilliard, 1817–20.

Blumenthal, M. "FDA Declares 259 OTC Ingredients Ineffective." *HerbalGram* 23 (1990): 32–33, 49.

Coulter, H. L. *Divided Legacy: A History of Schism in Medical Thought.* Vol. 3, *Science and Ethics in American Medicine: 1800–1914.* Washington, D.C.: McGrath Publishing Company, 1973.

Cragg, G., S. A. Schepartz, M. Suffness, and M. R. Grever. "The Taxol Supply Crisis. New NCI Policies for Handling the Large-Scale Production of Novel Natural Products Anticancer and Anti-HIV Agents." *Journal of Natural Products* 56, no. 10 (1993): 1657–68.

Eisenberg, D., R. C. Kessler, C. Foster, F. E. Norlock, D. R. Calkins, and T. L. Delbanco. "Unconventional Medicine in the United States: Prevalence, Costs, and Patterns of Use." *The New England Journal of Medicine* 328, no. 4 (1993): 246–52.

Farnsworth, N. R., O. Akerele, A. S. Bingel, D. D. Soejarto, and Z. G. Guo. "Medicinal Plants in Therapy." *Bulletin of the World Health Organization* 63, no. 6 (1985): 965–81.

Farnsworth, N. R. and D. D. Soejarto. "Potential Consequence of Plant Extinction in the United States on the Current and Future Availability of Prescription Drugs." *Economic Botany* 39, no. 3 (1985): 231–40.

Flexner, A. *Medical Education in the United States and Canada.* Bulletin no. 4. New York: Carnegie Endowment, 1910.

Ford, R. I. "Ethnobotany: Historical Diversity and Synthesis." *Anthropological Papers of the Museum of Anthropology of the University of Michigan* 67 (1978): 33-49.

Foster, S. "Goldenseal." *Hydrastis canadensis. Botanical Series, No. 309.* Austin, Texas: American Botanical Council, 1991.

—. "American Ginseng." *Panax quinquefolius. Botanical Series, No. 308.* Austin, Texas: American Botanical Council, 1991.

—. "Medicinal Plant Conservation and Genetic Resources: Examples from the Temperate Northern Hemisphere." *Acta Horticulturae* 330 (1993): 67–74.

Foster, S., and J. A. Duke. *A Field Guide to Medicinal Plants: Eastern and Central North America.* Peterson Field Guide Series no. 40. Boston, Massachusetts: Houghton Mifflin Co., 1990.

Hartwell, J. "Types of Anticancer Agents Isolated from Plants." *Cancer Treatment Reports* 60, no. 8 (1976): 1031–68.

Kartesz, J. T. *A Synonymized Checklist of the Vascular Flora of the United States, Canada, and Greenland.* 2 vols. 2d ed. Portland, Oregon: Timber Press, 1994.

Lloyd, J. U. "Materia Medica Americana: An Historical Review." Reprinted from the Century Issues of the *American Druggist and Pharmaceutical Record,* 25 March 1900. New York: n.p., 1900.

—. "A Review of the Principal Events in American Medicine." Reprinted from the *Eclectic Medical Journal,* February, April, June, and August 1926. n.p.: n.p., 1926.

Lloyd, J. U. and C. G. Lloyd. *Drugs and Medicines of North America.* Vol. 1, *Ranunculaceae.* Cincinnati, Ohio: J. U. and C. G. Lloyd, 1884.

Martin, S. "Unlabelled 'Drugs' As U.S. Health Policy: The Case for Allowing Health Claims on Medicinal Herb Labels: Canada Provides a Model for Reform." *Arizona Journal of International and Comparative Law* 9, no. 2 (1992): 545–92.

Moerman, D. E. *Medicinal Plants of Native America.* 2 vols. Technical Reports, No. 19, Research Reports in Ethnobotany, Contribution 2. Ann Arbor: University of Michigan Museum of Anthropology, 1986.

Piesch, R. F. and V. P. Wyant. "Intensive Cultivation of Yew Species: Weyerhaeuser's Contribution to the Taxol Supply Dilemma," *International Yew Resources Conference. Yew (Taxus) Conservation Biology and Interactions.* Edited by C. R. Temple. Berkeley, California, Native Yew Conservation Council, 1993, p. 27.

Schepartz, S. A. "History of the National Cancer Institute and the Plant Screening Program." *Cancer Treatment Reports* 60, no. 8 (1976): 975–78.

Tyler, V. E. "Plight of Plant/Drug Research in the United States Today." *Economic Botany* 33, no. 4 (1979): 377–83.

—. "Plant Drugs in the Twenty-First Century." *Economic Botany* 40, no. 3 (1986): 279–88.

—. "The Herbal Regulatory Dilemma: A Proposed Solution." Paper presented at the Drugs Directorate Seminar Series, Health Protection Branch, Health and Welfare Canada, Ottawa, Canada, October 1989.

—. *The Honest Herbal.* 3rd ed. Binghamton, New York: Pharmaceutical Products Press, 1993.

—. *Herbs of Choice: The Therapeutic Use of Phytomedicinals.* Binghamton, New York: Pharmaceutical Products Press, 1994.

Author

In introducing Steven Foster, Harvard University botanist Shiu Ying Hu wrote, "Our conversation reminded me of something that Confucius said two thousand years ago. 'In any company of three persons, there must be one who can be my teacher' . . . I found in Steven Foster a teacher who could share a profound knowledge of economic botany, particularly in the cultivation and uses of herbs."

Steven Foster is an author, photographer, and consultant in medical and aromatic plants. He began his career in 1974 in the Sabbathday Lake, Maine, Shaker Community's herb department, America's oldest herb business, dating from 1799. There he established three acres of production gardens and managed seventeen hundred acres for the commercial harvest of botanicals.

Foster is the author or coauthor of:

with Dr. A. Y. Leung, *Encyclopedia of Common Natural Ingredients Used in Foods, Drugs, and Cosmetics* (John Wiley & Sons, 1995)

with Roger Caras, *A Field Guide to Venomous Animals and Poisonous Plants of North America Exclusive of Mexico* (Houghton Mifflin Co., 1994)

Herbal Renaissance (Gibbs Smith Publisher, 1993)

with Professor Yue Chongxi, *Herbal Emissaries—Bringing Chinese Herbs to the West* (Healing Arts Press, 1992)

Echinacea—Nature's Immune Enhancer (Healing Arts Press, 1991)

with Dr. James A. Duke, *A Field Guide to Medicinal Plants: Eastern and Central North America* (Houghton Mifflin Co., 1990)

Foster is also the author of over four hundred photo-illustrated articles in popular, trade, and scientific journals. He serves as associate editor for *Herbal-Gram*, the *Business of Herbs*, and the *Journal of Herbs, Spices, and Medicinal Plants*, and as editor of *Botanical & Herb Reviews*.

Foster, who lives in Fayetteville, Arkansas, offers for commercial and editorial licensing America's largest stock photo library—over twenty-five thousand images—of medicinal and aromatic plants.